FLORA OF TROPICAL EAST AFRICA

SPHENOCLEACEAE

H. K. Airy Shaw

Annual glabrous hygrophytic herbs. Stem erect or decumbent, somewhat succulent and spongy. Leaves spirally arranged, simple, entire, exstipulate. Inflorescences terminal, densely spicate, acropetal. Flowers small, bisexual, regular, subtended by a bract and 2 bracteoles. Calyx-tube adnate to the ovary; segments 5, shortly connate, imbricate, connivent, persistent. Corolla campanulate-urceolate, perigynous, caducous; lobes 5, imbricate. Stamens 5, alternipetalous, inserted on the corolla-tube; filaments very short; anthers rounded, 2-thecous, dehiscing longitudinally. Ovary semi-inferior, 2-locular; style short; stigma capitate, slightly 2-lobed; ovules very numerous, anatropous, attached to large spongy stipitate axile placentas. Fruit a circumscissile capsule (pyxidium), cuneate-obconic, 2-locular, membranous. Seeds very numerous, minute, oblong; testa irregularly plicate-costate; albumen scanty or none; embryo axile, straight, subterete.

An isolated group, probably marginally related to the *Centrospermae*, e.g. *Phytolacca* (cf. habit, anatomy), and perhaps also to the *Primulaceae* (cf. capsule). Included by Bentham & Hooker and Engler & Prantl in *Campanulaceae*, probably due to convergence of superficial technical characters.

The family contains the following genus only.

SPHENOCLEA

Gaertn., De Fruct. 1: 113, t. 24/5 (1788); Baill., Hist. Pl. 8: 327, 361, figs. 158–161 (1886); Schönland in E. & P. Pf. IV. 5: 61, fig. 38 (1889); *nom. conserv.*

Characters of the family.

The only other species, *S. dalzielii* N.E.Br., occurs in West Africa from Senegal to the Central African Republic.

S. zeylanica *Gaertn.*, De Fruct. 1: 113, t. 24/5 (1788); W. B. Hemsl. in F.T.A. 3: 481 (1877); P.O.A. C: 400 (1895); F.P.N.A. 2: 405 (1947); Airy Shaw in Fl. Males., ser. 1, 4: 27, fig. 1 (1948); F.P.S. 3: 71, fig. 13 (1956); F.W.T.A., ed. 2, 2: 307, fig. 272 (1963); Roessler in Prodr. Fl. Südwestafr., 138, Sphenocleaceae (1966). Type: Ceylon, collector unknown (L, holo.)

Roots numerous, long, cord-like. Stem hollow, up to 150 cm. tall, often much branched. Leaf-blade oblong to lanceolate-oblong, attenuate at both ends, acute or obtuse, 2·5–12·5 cm. long; petiole up to 3 cm. long. Spikes cylindric, up to 7·5 cm. long, narrowed at apex; peduncle up to 8 cm. long. Bracts and bracteoles spatulate, the tips arched over the flowers except during anthesis. Flowers densely crowded, though characteristically only 1 or 2 open at a time, rhomboid or hexagonal by compression, sessile, wedge-shaped below, attached longitudinally to the rhachis by a linear base. Calyx-segments

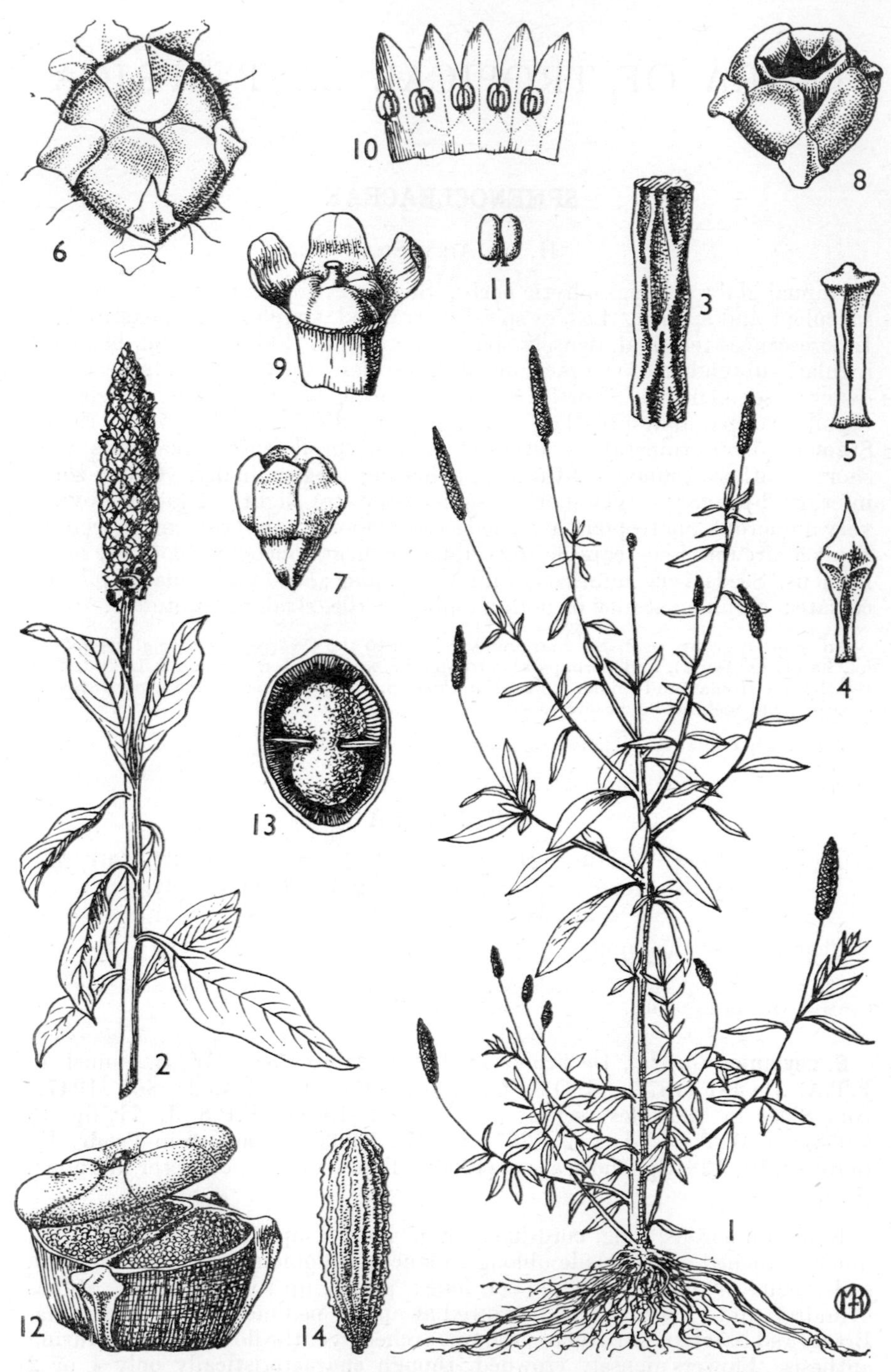

FIG. 1. *SPHENOCLEA ZEYLANICA*—**1,** habit, $\times \frac{1}{6}$; **2,** part of flowering branch, $\times \frac{2}{3}$; **3,** rhachis of inflorescence, showing scars left by fallen capsules, × 4; **4,** bract, × 4; **5,** bracteole, × 4; **6,** flower-bud, apical view, × 4; **7,** bud (from another plant) beginning to open, side view, × 4; **8,** flower, showing opening corolla, oblique view, × 4; **9,** gynoecium and calyx (with two sepals removed), showing cuneate base, × 4; **10,** corolla, opened out, × 4; **11,** stamen, × 8; **12,** fruit, partially dehisced, × 4; **13,** transverse section of fruit, × 4; **14,** seed, × 40. 1, 2, 14, from *Milne-Redhead & Taylor* 7463; 3–6, 12, 13, from *Jones* in *FHI* 18808; 7–11, from *Deighton* 132a.

deltoid-semicircular, obtuse, ultimately slightly accrescent and connivent. Corolla whitish, pinkish or purplish, 2·5–4 mm. long; segments ovate-triangular, obtuse or acute, united about half-way, connivent. Filaments slightly dilated at base. Ovary obovoid, 2·5 mm. long, apex broad, free, truncate. Capsule 4–5 mm. diameter, dehiscing below the calyx-segments which fall with the lid, leaving the scarious base persistent on the rhachis. Seeds yellowish-brown, 0·5 mm. long. Fig. 1.

UGANDA. Teso District: Serere, Dec. 1931, *Chandler* 131!; Busoga District: Bugwere, Oct. 1933, *H. B. Johnston* 603!

TANGANYIKA. Lushoto District: Mkomazi, 23 Apr. 1934, *Greenway* 3957!; Dodoma District: 27 km. W. of Itigi, 15 Apr. 1964, *Greenway & Polhill* 11570!; Rufiji District: Utete, by Rufiji R., 2 Dec. 1955, *Milne-Redhead & Taylor* 7463!

ZANZIBAR. Zanzibar I., Mwera, 22 July 1960, *Faulkner* 2663! & Ndagaa–Kisimbani, 2 Oct. 1963, *Faulkner* 3283!; Pemba I., Chake-Chake, 18 Aug. 1929, *Vaughan* 523!

DISTR. **U3**; **T1**, 3, 5, 6; **Z**; **P**; widespread in tropical Africa (excluding the NE. Horn), extending south to the Transvaal; also in Madagascar; widespread (but probably introduced) in tropical Asia and America

HAB. In and near pools, swamps, streamsides, periodically inundated depressions, ditches, irrigation channels, and wet places generally; 0–1250 m.

NOTE. The alleged presence of laticiferous elements in the stem of *S. zeylanica* is so far unconfirmed.

INDEX TO SPHENOCLEACEAE